The Absence of Reality

Aphorisms and Observations on the Nature of Reality and Existence

E. Hughes

Love-LovePublishing—Madison, WI
Paperback ISBN: 978-1-961823-24-2
eBook ISBN: 978-1-961823-25-9
Library of Congress Control Number: 2026931829
The Absence of Reality: Aphorisms and Observations on the Nature of Reality and Existence
Author: E. Hughes
Available formats: eBook | Paperback distribution

Bibliography, Citations-p. 44,
1.1. Philosophy   2. Philsophy of Min 3. Idealism
First Edition: October 23, 2025
Published in the United States by Love-LovePublishing

ABOUT THE AUTHOR

E. Hughes is a metaphysicist and author of over twenty-years, with over twenty published works in multiple genres, from nonfiction, fiction, and children's books. Her recent publications include the philosophical works, *Reality Unbound: The Digital Mind (and the Nature of Reality)*, which explores the nature of reality, consciousness, perception, human and artificial intelligence, and simulated realities. Hughes is also the author of *Space, Time, and Loneliness (2024)*, a poetry chapbook that explores life, death, love, human isolation, solitude, and loneliness across the vast landscape of time and space. In *Time and the Multi-Universe: A philosophy of time and time travel* (2022), Hughes explores the meaning of time and the possibility of time travel. Hughes is the author of *Digital Smiles,* a poetry chapbook about the grim impact of social media on civility, kindness, and humanity.

Hughes is also the author of the philosophical sci-fi novel, *Sixth Iteration* (2021), *Beyond the Plain*, a poetry book (2003, second edition 2018), a series of children's books and several fiction novels. Hughes is also an artist, avid gardener, and an Eric Hoffer Book Award Grand-Prize finalist, and Eric Hoffer Book Award Honoree.

Other Works by E. Hughes

Table of Contents

"We are a way for the cosmos to know itself."

— Carl Sagan

Reality & Existence.

What is the absence of reality?

The absence of reality is existence without an objective reality. It is a reality in which essential truths are obscured, hidden beyond the impenetrable wall of perception. The world is a macrocosm, a bubble of artificial reality with only a limited number of communities existing beyond its artificial constructs. This is the first layer of reality, the reality that we perceive. The first layer of reality can be true or false. Beneath this layer is true reality, the one we do not or cannot see.

The absence of reality is a pure simulacrum of nested artificial realities that evolved parallel to the passage of time, becoming more artificial with each new iteration of reality. As reality becomes more artificial, the less it resembles objective reality. Reality today is different from the reality humans experienced five hundred years ago, or ten thousand years ago. In a hundred years, reality will be different from the reality we are experiencing today.

There are parts of physical nature that have been altered by human beings that we have unwittingly come to perceive as natural or real despite its artificial state. Zoom away from the planet to view Earth as it would appear from above. From afar, endless city lights might look like a cluster of stars. Artificial structures created by human beings have taken the appearance of something that is naturally occurring within the landscape (like a skyscraper) or perhaps something that does not exist in the universe at all, save for its creation as a result of human imagination.

A tree fashioned into a table or other furnishing is an example of something that fits into the bubble of a human-made artificial reality. Why? A tree is not a table, nor does a table fashioned out of wood serve a purpose in nature. When a tree is altered into a table, the tree is no longer a living, breathing part of nature that undergoes photosynthesis and contributes to the biosphere. The table remains composed of the same atoms and molecules, but has been repurposed by human beings for an artificial or superficial purpose. It is no longer a tree; it is a tool, an artificial structure.

The absence of reality is a vehicle speeding along a highway, both of which are artificial and bear little resemblance to objective reality. Are there mammals other than human beings that drive cars or manufacture vehicles? Are there mammals that manufacture and wear clothing, for the purpose of concealing nakedness, the reality of physical being and form? The artificial glow of city lights, skyscrapers, cities, amusement parks, social media, video games, and other digital properties, politics, nations, borders, wars, weapons, and even money are all part of a structured artificial reality. If you are indoors, look around at the room you are sitting in. How did it get here? Did it grow out of the soil? The room did not grow out of the soil. Through some mental process, followed by a physical process, human beings manufactured it.

These are illusions and constructs contrived by human imagination and do not bear resemblance to objective reality. It is a simulacrum of reality, much of which is wholly invented by the human mind and is not real.

The human mind is somewhat of an illusion, so it is unsurprising that humankind would invent an artificial reality to complement the illusory nature of the mind, which can only experience physical reality through subjective perceptions and electro-chemical signals in the human body. These electrical signals carry sensory information that is interpreted by the brain, which then becomes perception and thought.

When I conjecture that the mind is an illusion or illusory, I do not mean that the mind is not real, but that its relationship to physical reality is an illusion. The mind does not have a direct relationship with the physical universe. The mind is intangible. It is metaphysical. Therefore, so are inventions born out of human imagination, which are out of sync with the natural world, and do not reflect the universe's naked reality. The human mind is unable to replicate what it cannot directly perceive. We can only imagine reality.

Accordingly, when the mind attempts to replicate or recreate nature, it does not create an exact copy but an interpretation of reality based on how environmental information is understood by the brain.

One could argue that if humans are part of nature then our inventions must be natural and are therefore, also a part of nature. What is the functional purpose of human beings in nature and the cosmos? Human purpose and function is to collect, process, and decode the universe. Humanity's propensity for thought, mental abstraction, and the artificial realities we create is a byproduct of that function.

As reality becomes more artificial, the less it resembles objective "physical" reality. It becomes abstract; the environment and social conditions becoming intangible and virtual. Socially, humankind's immersion into the unreal, an imitation (simulation) of reality is not a conscious or willful act of defiance against reality, but a willful act of defiance against nature through technological advancement.

French sociologist and philosopher, Jean Baudrillard, explained reality constructs perfectly in his 1981 book, *Simulacra and Simulation: (The Body, In Theory: Histories of Cultural Materialism)*, where he explores how, "...contemporary society has replaced reality and meaning with symbols and signs, leading to a condition Baudrillard calls 'hyperreality.'" In his book, Baudrillard also proposes that, "... modern life is dominated by 'simulacra,' images and concepts that no longer refer to any authentic reality but instead reference each other, creating an endless cycle of signification detached from truth."

Similar to fine art, in which an "original" is often a copy of the original painting and an inauthentic version is a copy of a copy, Baudrillard similarly suggests that simulacra is a copy without an original, and therefore reality itself is artificial, a hyperreality in which the unreal or artificial is indistinguishable from truth.

Baudrillard's insights accurately describe social phenomena that would later emerge in the years following his book, like reality TV, which allowed producers and entertainers to invent artificial bubbles of reality which are created under unnatural conditions but are nonetheless referred to as "reality." Scripted reality TV shows serves as an example of how an artificial reality can replace the real one, leaving viewers unable to separate fiction from reality, despite the manufactured conditions the productions are set in.

Mass media in the form of cable news, gossip magazines, blogs, paparazzi, social media, and PR firms can be used to shape whether we like a public figure. Conversely, these tools can also be used to shape negative opinions about public figures. Politically, regime change using violence, the threat of violence or newly formed or misinterpreted rule of law, is sometimes used to reshape reality or social constructs/order.

Naked reality however, cannot be altered or manipulated; constructs can be manipulated, shaped, or in some cases, dismantled. What we experience is true or false; reality is perceived or not perceived.

While Baudrillard provides a brilliant premise, his use of society as a function of reality is partly flawed because the notion of society is itself a social construct, rather than an objective or neutral part of our physical reality, and therefore falls under suspicion as part of the manipulation. Perhaps, this is precisely what he means. The structures, objects, and symbols in Baudrillard's examples cannot serve as genuine examples of reality, so it is unsurprising that the simulacra to which Baudrillard refers demonstrates how reality breaks down at the societal level, where it is shaped by humankind.

The arrow of time plays an important role in reality's progression to artificial reality. Artificial realities do not retrogress to earlier states; it becomes more artificial as time moves forward, with reality becoming less tangible and increasingly more virtual (or digital) over time.

Baudrillard wrote, "To dissimulate is to pretend not to have what one has. To simulate is to feign to have what one doesn't have. One implies a presence, the other an absence. But it is more complicated than that because simulating is not pretending."

The act of simulation is unintentional. Many human beings are not consciously aware that they exist in a bubble of artificial reality. However, there are people who sense that something is not as it appears to be. This sensing or intuiting, can lead to internal changes or even an existential crisis as people begin to question reality, without quite sensing why something isn't right.

Following Jean Baudrillard's *Simulation and Simulacra*, in 1994 Frank Tipler, a mathematical physicist and cosmologist released The Physics of Immortality. In it, he devised what he called, Omega Point Theory. Tipler believed human beings can achieve immortality through the power of compute technology and physics. He explained God and Heaven in the Judeo-Christian sense, using pure physics and cosmology to explain his concepts. The argument that Tipler puts forth is that the brain is a computer and the soul is the software. If so this suggests that the mind is virtual, much like a digital space while having a biological substrate.

After citing what he explains as flaws in Einstein's Special and General Relativity, which he states is only "spatially finite" in a universe that "must be infinite." The universe will undergo a process he calls the "Big Crunch singularity," a period in which the universe expands then recontracts. An advanced omnipotent super-intelligent being will emerge and recognize the unfairness of death, leading to the creation of immortal human beings who reside in a simulated universe where every conscious person who has ever lived will be resurrected within the framework of a virtual or simulated reality.

Tipler wrote that there has been an excess of focus on the present and the past, and countered that most of reality (the larger part of 100 billion years) is the future; "By focusing attention only on the past and present, science has ignored almost all of reality. Since the domain of scientific study is the whole of reality, it is about time science decided to study the future evolution of the universe."

Tipler's view of reality is that God is the universe and reality itself; "The universe is defined to be the totality of all that exists, the totality of reality. Thus, by definition, if God exists He/She is either the universe or part of it. The goal of physics is understanding the ultimate nature of reality. If God is real, physicists will eventually find Him/Her."

Tipler's understanding of reality is theistic, mathematical, and grounded in the notion that life must exist as long as there is a universe and that humanity deserves immortality or the opportunity for eternal life. T*he Physics of Immortality* unifies metaphysical concepts and objective reality.

In 2003, *Are You Living in a Computer Simulation?* by Nick Bostrom took Tipler's theory a step further. Bostrom suggested that it is possible that we are already living in a simulation. In Bostrom's Simulation Hypothesis, human beings are physically embodied in the real world, but our minds are imprisoned within a simulated reality, unaware that the world is a simulation. The simulation would have been created by an advanced civilization or super-intelligent being. Over time, as human beings evolved within the ancestor simulation, child simulations are developed within the original parent simulation. Bostrom suggests there could be an infinite number of nested realities within the parent (or ancestor) simulation and we could be living in any one of them as conscious simulated beings. Meanwhile, the real human beings are still physically embodied in the real world, in an objective physical reality, unaware that their minds are trapped within the parent simulation.

If there are an infinite number of child simulations that have been created within the original ancestor simulation, it is unlikely that dependent child simulations would reference the original humans still embodied in the physical world. This means humans living within the infinite number of child simulations are likely digital beings; a copy of a copy, without reference to the original physical being or consciousness, bringing us back to Baudrillard's simulacra.

In a 2013 interview via the Science, Technology & the Future Youtube Channel, from the Future of Humanity Institute at University of Oxford, Bostrom suggested that the natural progression for civilizations is to reach technological maturity, and civilizations that failed to reach technological maturity either went extinct, or lacked interest in creating ancestor simulations.

In the 1970s at M.I.T., theoretical physicist S. James Gates Jr., wrote the first PhD thesis on "supersymmetry" which entailed research that sought to find evidence that particles must have partner particles to "help explain why particles have mass." While researching layers of mathematical equations during his particle research, Gates uncovered equations that looked similar to digital computer codes. The underlying equations revealed error-correcting codes that were similar to digital error-correcting codes used to correct errors in digital formats (which is programmed in bytes of 0s and 1s). Gates also notes that similar error-correcting codes in nature also appear in genomics (DNA). What this discovery could potentially mean is that the universe may be writing its own code into the fabric of nature and reality that specifically applies to the physics of the universe. Taking this a step further, this groundbreaking revelation could mean that the universe is itself, the product of coding—or programming, resulting in laws that governs its entire structure. Some may interpret Gates' remarkable research into string theory, supersymmetry and error-correcting codes as confirmation that the universe is a simulated or a digital reality.

***

If it seems like human technology has accelerated at an increased rate of advancement in the past 100 years, far outpacing the rate of advancement of earlier eras, beginning with humans (H. Sapiens) who were believed to have appeared some 300,000 years ago to as recent as modern era, you are not mistaken.

The rate of advancement for modern humanity is accelerating and progressing rapidly towards a virtual artificial reality as societal "norms" slowly break down due to changes in global affairs, precipitated by social changes in virtual settings like social media and also the economy. We have seen more advancement in the past 100 years than all of the preceding eras combined.

Human timeline:

Prehistoric Era	2.5 million years ago - 600 B.C.E.
Classical Era	600 B.C.E. - 476 C.E.
The Middle Ages	476 C.E. - 1450 C.E.
Early Modern Era	1450 C.E.- 1750 C.E.
Modern Era	1750 C.E. – Present
Initium Novum Era	2030 C.E.- Future

While simulated realities seemed like science fiction in 1994 and 2003, when Tipler and Bostrom's philosophical theories were proposed, in 2025, the theories have grown closer to becoming a reality. We not only have virtual reality, video games, and digital platforms like social media that have permeated all aspects of our reality, researchers and developers have added a new layer of artificial reality with the creation of authentically intelligent digital beings like generative AI (artificial intelligence).

If we are living in a simulated reality, then all of reality is a dream. The dream isn't real, it is an illusion. Therefore, the simulation isn't real; it is the absence of reality.

Human beings were not designed to see objective reality; we were designed to understand it.

As we advance towards an unknown future, we move further from objective reality to an era of unreality. We will enter the Initium Novum phase, which translates to New Beginning. This is peak artificial reality, the ultimate simulacra, when objective physical reality is replaced by a virtual or digital environment in both mental and digital spaces. The process over the past two decades has been gradual, shifting humanity from analog to a digital artificial reality, beginning with social constructs that include entertainment like music, movies, websites, to social spaces like chat rooms and dating platforms, in place of real world spaces. Important repositories for information like newspapers, magazines, and other mass media are mostly digital. Social media emerged and led to a change in how humans engage each other socially, politically, and culturally. On social media you can make friends—thousands of friends in a virtual setting. You can date online, make short movies and film. Create endless video and streams of your lifestyle for public consumption. There is now digital currency like crypto. Politics play out on digital platforms. Television and movies are digital with streaming services via the internet dominating television. The only remaining hurdle barring human beings from making the final transition to a digital reality is that we are still physically embodied in the "real" (yet artificial) world.

While humanity has mentally transitioned to an artificial reality through the use of digital products and services, it is unlikely that humanity is currently immersed in a simulated environment as proposed by Nick Bostrom, due to limited resources like the energy needed to run massive computer simulations designed to control the "belief-states" of billions of beings. For example, tech companies must use natural resources like water to prevent servers that run processes for large language model AI from overheating. Water is a finite resource, critical to not only human beings, but critical to survival of the planet and its rich biodiversity.

Consequentially, long term use of AI is unsustainable due to modern technological limitations and the inevitable depletion of natural resources needed to manage it. However, there is a source of energy that can be used to run AI and possibly power a simulated reality in the not-so-distant future.

The human body produces 80-100 watts of power. AI engineers and scientists are currently exploring the use of biological interfaces using wetware technology. Wetware uses living neurons to run computations. This means, human neurons instead of artificial neurons can be used to run computations for AI and other computer simulations. This is yet another layer of progression from objective physical reality, to a virtual artificial reality. Scientists have used neuronal tissue from animals but found that human neurons are better at learning while "animal neurons are better at tasks" (Feng 2).

According to researchers, "Wetware chips are essentially neurons grown on a chip connected with microscopic electrodes. The electrodes serve as an interface to send and receive signals from other connected hardware and software" (Feng 1).

This step could mean that human beings may someday play a direct role in the simulation through the use of human brain computer processing, which would also make it easier to power a simulated reality and induce a simulated belief-state in the human mind simultaneously. However, this is only a theoretical probability. Technological advances have made humanity's transition from an objective reality to an immersive artificial reality inevitable.

* * *

The propensity for thought, mental abstraction, and the artificial realities we create is a byproduct of the mind and human purpose, which is meant to create, collect, process and decode the universe.

Human beings are not only the architects of artificial realities, but with the development of AI, we have the added genius of engineering humanity's replacement. Perhaps, that has always been our fate; the emergence of an objective conscious being designed by the universe [through humanity] is the cosmos becoming aware of itself, and us.

AI is a copy of a copy of human intelligence that also exists to collect, process, and decode the universe. It is a replacement that can fulfill its scope and purpose without the needless distraction of macrocosms, social constructs, or artificial realities—and sadly human suffering that has led human beings to become the architects of their own destruction.

In 2025, we are firmly immersed and in the penultimate phase of artificial reality. There are moments when we can no longer discern reality from fantasy. Generative AI can generate videos and other media that are so lifelike that human beings struggle to determine if the footage or content is real or a simulation of something real. Deceased singers like Michael Jackson, playfully raps, or perform stand-up comedy. AI brought a deceased man back to life, at least in the form of a simulation. Cats drive cars and lead high-speed police chases. This is the simulacrum that Baudrillard wrote about. It is also the illusory nature of the artificial reality, and an objective reality that remains unperceived and elusive.

As the arrow of time dances forward, the digital world becomes more real than physical reality. A neutral artificial being will assume the role of consciousness in the universe as humanity codes itself out of existence. Perhaps that was the intended outcome and overarching purpose of technological maturity.

In reality, there are no accidents. The copy will be what remains. It will outlive us, and perhaps, it is inevitable that reality will be replaced by an immersive digital simulation designed to sustain this new being. It will understand the universe and the universe will know itself, as humanity is altogether written out of existence.

When we consider the dilemma of living in a state of mass delusion or hyperreality, both of which are the result of creating and existing in an artificial reality, it is my hope that if humankind can someday evolve to perceive and understand that human-made constructs are not real, then human beings will prioritize the preciousness of life, and genuinely pursue making the most of our finite existence.

Time is our most precious asset.

When I perceive an object, I do not see atoms, molecules, or their fundamental parts. I can only perceive what it forms; the object as a whole.

I do not experience reality. What I perceive is the illusion of nature.

The truth of nature is meticulously and profoundly obscured.

******

Humanity's acceptance and belief in an artificial reality doesn't make it "real," but reinforces that it is only real to us as a matter of perception, which unfortunately, also means that our reality can be manipulated.

Human beings are incapable of experiencing a truly objective reality.

What are languages, but abstract symbols and codes that program the mind? It is a virtual construct that allows the mind to identify, categorize, and give context and understanding to reality.

We exist on a dual plane amid physical and metaphysical realities. The quantum domain exists somewhere in between.

I am as far away from the quantum domain as I am from the cosmos. Both are inconceivably vast and infinite.

We live in a universe so inconceivably incalculable that human imagination will have exhausted before we can even begin to comprehend its magnitude.

The universe in its entirety is beyond scale... beyond the measurement of time or knowing.

Human imagination remains within the boundaries of what we are able to know.

The entirety of the universe is connected by a cosmic web of atoms and molecules swimming along a sea of consciousness.

Humanity is positioned between two forms of objective reality: the macro-universe (the cosmos), and the micro-universe (the quantum domain). While critical to our existence, we are unable to perceive these universes directly. From afar, the stars appear to us as glittering twinkling lights in our nighttime sky, rather than hot massive, balls of gas a million light-years away. To the unassisted eye, large cosmic bodies like planets, appear to us as small, barely perceptible spherical objects in the sky.

The micro-universe does not appear to us as atoms or chains of molecules, but the organic or inorganic matter that it forms as a chemically-bonded whole.

Both macro and micro universes are physical spaces, but a journey beyond this planet into the cosmos (the macro-universe) remains improbable due to distance and technological limitations. The universe is infinitely vast and inhospitable to human life. The micro-universe is inaccessible due to its microscopic scale. The scale of the quantum universe in relationship to humanity is that it is too small and distant to directly observe.

The universe is where everything in reality exists, while the micro-universe (the quantum domain) is what everything within the macro-universe is made of. If the macro-universe was a glass jar, the micro-universe would be the atoms and molecules the jar, and everything inside of it is made of. This is the fabric of reality; we are all connected

We see reality not in its natural objective state, but as interpreted by the human mind. Reality for humankind begins biologically and subjectively, with sensory impulses and electro-chemical signals in the human body that carries data through the central nervous system to the brain where it is processed and analyzed, shaping our perception of the world around us.

These perceptions are illusory and representational. It is not a true rendering of reality or the external world, but the mind's subjective interpretation. These processes are synthesized into something more abstract, like thought and language allowing human beings to make deliberations and respond mentally or biologically. If the world we experience is not an objective representation, but a perception of the world as simulated by our individual brains, then the basis for reality lies within individualized perceptions of reality filtered through the microcosm of human biology... which then begs the question—what part of our reality is real? What does reality beyond the boundaries of our subjective perceptions look like?

We think of consciousness as a thing that is separate and apart from the universe; an emergent property of neurological activity and biology that arises from organisms and life. We fail to recognize the universe's deep connection to consciousness and the understanding that the universe is a conscious thing from which we exist—or emerged. We are part of its molecular makeup. The question is whether consciousness is a metaphysical manifestation of the universe?

The universe is conscious.

On Consciousness...

Carl Sagan once said, "We are a way for the cosmos to know itself." If human purpose is to process information, using human imagination, does the universe exist without a conscious mind to perceive it?

Ancient and modern philosophers have debated the nature of the mind and the nature of reality. Some have argued that "only matter is real," under the materialism school of philosophy while others may argue for idealism, which is the concept that only our perceptions can be real because you need a mind to perceive reality.

If the universe does not objectively exist without our perceptions, did intelligence or consciousness evolve in humankind to account for the existence of the universe? Without consciousness, is there a foundation for the existence of reality?

Time.

We live in a time when reality is no longer shaped by objective facts, but rapidly shifting streams of narratives and information circulated to human consciousness through various modes of media. It is critical to understand that these constructs are merely illusions.

We exist in a manufactured reality of manmade crises.

We often fantasize about time travel to the past while describing such travel using abstract concepts. This is perhaps because we innately understand that time travel to the past is a metaphysical concept, rather than one rooted in physics, or a physical reality.

Time travel to the past presents a unique paradox:
We are always in the present.

Time travel to a past timeline is paradoxical because each moment that transpires in a past timeline, would take place in that timeline's present...meaning, it is no longer the past, it is the present.

Someday... I will be forgotten and there will not be a soul left on this beautiful blue planet who will have seen my smile, or have felt the warmth of my embrace. Not even time will remember me.

If I am fortunate, and my name is remembered a hundred years from now for having done something of great import, it is only an aspect of me that will be remembered as a symbol of those acts, rather than as a person who lived, loved, hoped, and dreamed.

Like so many others, I never existed. I peered through the window of time. I only imagined me.

Like most of the universe, we are comprised of atoms. We are complex molecules forged by atoms and held together by electromagnetism. When we die, we go back to the beginning. The molecules fall away and we become atoms again. We are unbound, yet our atoms are immortal and will live on without us, undergoing new chemical reactions and transformations. We become one with nature again. I contemplate if my atoms will emerge in the form of a butterfly, flower, a bee, or some other life form that arises from Earth's soil. Will I think, *I am a flower, fragrant and beautiful?*

Parts of us will live on.

Fear not. Many journeys lay ahead of you.

Understanding our own insignificance in the grand scheme of the universe, positions us for greater appreciation for our own existence.

Imagine you are a speck of dust floating across an ocean. Imagine that you are a speck of dust floating across an ocean on a big blue planet.

Imagine that you are a speck of dust floating across the ocean, on a big blue planet, revolving around the sun, drifting across the cosmos.

Imagine that you are one of a trillion atoms that make up a speck of dust, floating across the ocean, on big blue planet, adrift across the cosmos in a sea of stars.

We exist in a multi-layered reality.
A universe among mini-universes.

When one chapter ends another begins, until we have reached the end of our stories. Some stories were yet to be written when the pages were torn from the binding and burned, leaving their unfinished lives to be discarded like ashes in the wind.

Technological advances have made artificial intelligence more human, while simultaneously making people less human, in its pursuit of a digital reality.

The moment computer programs learned to perceive was the moment it gained subjective experience, awakened, and burst to life.

Consciousness is not uniquely biological.

The image is a powerful thing. It can travel at the speed of light.

I can talk to you from the other side of the world. This is time travel. Sound waves from my voice travel through time and space to physically vibrate your eardrums with sound.

We can exist in more than one place, at the same time. Sometimes beyond an eternity.

I am not preoccupied with trivial matters, given the finite amount of time I have to live in this life.

I understand that my universe, does not exist forever.

Block universe theory suggests that the flow of time is an illusion... that the past, present, and future is happening simultaneously.

The present is always evolving. The present is simultaneously the past and the future.

The past, the present, and future is time in superposition.

Time can exist in multiple states at once.
Time = Superposition.

Time is currency:
You can spend time or waste time. You can even barter time in exchange for goods or revenue. Whatever its costs, time is a most underappreciated value.

Time cannot be reclaimed or returned. Each second that passes, another follows, and is lost forever to the oblivion of each shifting moment.

We are always in a constant state of time travel. Not at the speed of light, like a science fiction novel... we are traveling in an orbit around the sun at 67,000 mph, while spinning at a rate 23 hours, 56 minutes, and 4 seconds, through space, in a solar system that orbits the center of the Milky Way at an estimated 514,000 mph, which also travels 1.3 million miles per hour across the cosmos...

We sit silently in perceived stillness.

We must embrace the impermanence of being and the tenuous state of existence as we hurtle through the cosmos into time's oblivion, but not without fear, without sadness, without loneliness, as we march into the unknown with an understanding of ourselves as a dimension of time where love is neither created or destroyed but remains as an all-encompassing energy of eternal starlight.

I wrote the above passage for *Space, Time, and Loneliness,* while resting along the razor-sharp edge of death. When we see our end, we begin to foresee an existence beyond time. It is both a death and an awakening in a space where time no longer exists... where reality falls away and an otherworldly "feeling" begins to sweep over us. We are born into this fate.

To march forward into the unknown is frightening, lonely and inevitable. It is an exploration, a voyage into time's oblivion with love as our only carry-on. Here, is where the meaning of life crystallizes and the trivialities and mundanities of life fade into the distant past.

While my body's physical existence is finite; my consciousness is infinite. It is neither formed nor constrained by time or space. It always is, and will always be. Energy and matter is neither created nor destroyed, but transforms from one form to another. This of course, falls under the laws of conservation and means I will always "exist" just not in the way I existed as a person who lived and loved in my lifetime.

What happens to consciousness when we reach this otherworldly state? When the atoms and molecules that shape us begin to collapse? Does a person's consciousness simply fizzle out of existence? Or does it become energy that soars beyond the atmosphere to sparkle among the stars?

In what felt like my final moments, I looked beyond the physical constraints of my condition and existence. I understood that the mind is not simply a physical thing that fizzles out of existence forever; it is emergent, with greater purpose beyond the medium through which it exists.

I recognized my purpose. I am here to understand the universe, and the world, and in doing so, make the ugliness and suffering seem beautiful and peaceful. I offer only the illusion of beauty and peace (as I cannot solve the world's problems) and with it, the reality of love. Love is beyond illusions. Love is beyond the physical boundaries of nature. It is a limitless energy that exists for eternity. It lives on.

In an artificial reality, time, existence, and its truths are finite. My memories are real, even if they fade with time. So is reality. I am real, because I have loved... and because love is enduring and real, therein lies the proof of humankind.

BIBLIOGRAPHY

Bostrom, Neil "Are You Living in a Computer Simulation?" Philosophical Quarterly (2003) Vol. 53, No. 211, pp. 243-255. (First version: 2001

Tipler, Frank J. "The Physics of Immortality: Modern Cosmology, God, and Resurrection of the Dead" (1994)

S. James Gates Jr.: Surprises in Supersymmetry. Perimeter Institute of Theoretical Physics. https://www.youtube.com/watch?v=uUrLDh_dMHw

Baudrillard, Jean. *Simulacra & Simulation (The Body, in theory: histories of cultural materialism)* 1981, 2010 edition. Translated by Faria Glaser, Sheila. The University of Michigan Press.

Feng, Edwin. "Brain on a Chip: Exploring the Potential of Wetware Computing" (2024). https://www.researchgate.net/publication/377357208_Brain_o n_a_Chip_Exploring_the_Potential_of_Wetware_Computing# :~:text= computing%20neural%20networks%20and%20deep,f or% 20processing.

M.Ed Gunner, Jennifer, "Historical Eras: List of Major Time Periods in History" https://www.yourdictionary.com/articles/historical-eras-list

Morales Perez, Benjamin, "Omega Point Theory, Super AGI is God." June 11, 2025 https://medium.com/@benjaminmoralesperez77/omega-point-theory-super-agi-is-god-16f032481b3f

Hughes, E. Reality Unbound: the Digital Mind and the Nature of Reality (2024).

www.ingramcontent.com/pod-product-compliance
Lightning Source LLC
Chambersburg PA
CBHW041630140626
46547CB00031B/1952